曾敏雄

2015, 02. 12.

風景　安靜

VOICE INSIDE

曾敏雄攝影集

Tseng Miin-Shyong
1997-2007

我一直想走到那寂寞的世界

四面光明　一處黑暗

四面黑暗　一處光明

在那個地方

思想與存在

獨自一人

風景　安靜

　　照片所具有的二度空間與聲音過濾特性，基本上應該是平面而安靜的，是隔一個「距離」的觀看再現。風景靜物所呈現的世界，更必然是無聲的，但實際上我們看到不少平汎、造作或俗艷的風景照，不時釋出嘰喳、嘀咕或瑣碎的訊息，一點都不安靜。

　　每一幅「風景」其實都自若、安靜地存在於它自己的特定時空，攝影者是否適時地進入它的界域或磁場，是否感知一種直覺的相互觸動，決定了對望之後的影像複製與再生。這樣的「風景」，不是因為你看見它，是它先看見你，召喚你按下快門。你只是偶然間一晃而過，它卻是久久、靜靜地倚立著，等候有心者看見。

　　初次看到曾敏雄鏡頭裡的風景，便為其中的單純的造型、細緻的質感與靜謐的氛圍所吸引。這些風景，與他之前的人像作品絕然不同。在他長年紀錄的人物攝影集裡，我們看到許多目光如炬的臉孔瞪視或遠望，許多人為的姿勢與打光，每一幅臉孔似乎都想訴說無盡的歲月情緒，直接、有力，但也略顯刻意與躁進。然而，這一批風景造像卻顯得如此自然、寧靜、致遠，有一種距離，有一種詩意與想像。

　　飄浮的煙塵、消隱的泡沫、斑剝的門牆、廢棄的鐵道、龜裂的土地、冬日的斜陽、肅靜的田野與沙灘 …… 曾敏雄的島內風景，靜默中隱含三個面向的關照：物象的質素、美學的訊息及人文的視野。有些寂寞，有些悽愴，但愈顯真實、內在與厚重。

　　就像美國畫家Georgia O'Keeffe在新墨西哥州找到屬於她的宗教，曾敏雄追隨她的腳步到這兒找到了一種「心鏡」的嚮往。透明的穹蒼、純淨的陽光、蕭瑟的曠野、無垠的沙灘、碑墓般的教堂、曙光乍現的湖濱、恆古的石丘與城堡 …… 一如幽靈般的「遠方」(The Faraway)。何其寂寥，何其聖潔與神秘，好像返回人類誕生的原始腹地。

　　即使曾敏雄的風景裡包舍的幾張人像與特寫，竟也安靜得叫人屏氣凝神。在孩童的眼神、少年八家將的擺示、老者的鬍鬚、告別式的遺照等所建構的氛圍中，彷彿進行的是一場無言的人生記憶，必須安靜，才能重重嵌印在腦海裡。

　　曾經有一篇短序中這樣說：「我一直想走到那寂寞的世界，四面光明，一處黑暗；四面黑暗，一處光明。在那個地方，思想與存在，獨自一人。」曾敏雄的內心風景，猶如那首他所嗜愛的修伯特名曲 --「冬之旅」一般，兀自在蕭澀的曠野中行走著，以孤寂、沉澱的心境，尋找時光的消逝跡痕，回歸歲月的永恆懷抱。

張照堂

Voice Inside

Two dimensional images and a voiceless attitude are characteristics of photography. It's a view or a presentation at a distance, so it should be flat and quiet.

For the same reason, the world of landscapes and still lives must be voiceless. However, we see many normal, overdone, affected and vulgar landscape photos which chatter noisily and send a fragmentary message. They're never quiet at all.

Every image exists in a still and self-composed space and time. Could the photographer insert himself into the image's domain and its magnetic field? Could he be aware of a mutual touch intuitively? All of them decide what is duplicated and reproduced after looking through each other's eyes. Such an image could not be seen; it looks at you first. It inspires you to press the shutter. You are merely a bystander passing accidentally. Nevertheless, an image is still and long-standing, waiting for an observant person to notice it.

At first sight, I was attracted by the simple shape of the landscape in Tseng's photos, as well as its exquisite quality and quiet atmosphere. These landscape photos are absolutely different from his previous portrait photos. In Tseng's portrait photos, there were many faces with torchlight-like wide open eyes staring at us or into the distance. There were many posed gestures and light settings. Each face seems to want to express its emotions. They were direct and strong, but also a little artificial and too hasty. In spite of this, these landscape photos are so natural, calm, and tranquil. Inside these works is a poetic imagination at a distance.

Floating dust and smoke; bubbles slipping away; a dappled wall and gate; obsolete rails; fissures in the ground; the stillness of the fields and sands... all of Tseng's landscape photos taken on this island imply three points of view silently. The first is the fine quality of his objects and images. The second is the aesthetic messages in his photograph. The last is his humanistic view. Even though they are a bit lonely or dismal, they bring out something intrinsic, truer and more immense.

Just like the American painter Georgia O'Keeffe, who found her own religion in New Mexico, Tseng followed her footsteps there to fulfill his hankering for "the mirror images of my mind." A limpid sky, the pellucid sunshine, stagnant wastes, endless sands, the tombstone-like church, a lakeside at dawn, hills and castles from time immemorial...like ghosts they come from "The Far away." Such silence, such holiness and mystery is like returning to our ancestors' wombs.

They are so quiet that you are hardly able to breathe in front of them. It seems that a voiceless memory of life is coming to you through the gaze of a child, from the pose of young Jiajiang, in the elder's beard, through the picture of the dead. An atmosphere is formed. We must be quiet in it's presence to let the images imprint on our minds.

Once a short preface said, "I want to walk straight toward the desolate world, in which every place was blighted but some places were dark; in which every place is absolutely dark but some places had light. In such a place, thoughts and existences are all alone." The inner sight of Tseng must be like his favorite song *Winterreise*, Franz Schubert's famous lieder, walking alone in chilly and stagnant wastes. With the solitude and disposition of his moods, he has been looking for the trails left behind space and time to send them back to the harbor of eternity.

Chang Chaotang

我仍舊做著一個攝影的夢

　　2005年，在攝影集【六十，六十】的自序中，寫下了【我做了一個攝影的夢】，大意是如果有一陣子無法出外拍照，那麼，我時常會做一個夢，夢中帶著相機出外拍攝，然而，就在碰上一個好場景或是一道絕妙的光線時，卻找不到要用的鏡頭或是裝不上底片......然後，就驚醒了!

　　2007年五月，我到美國的新墨西哥州流浪，追尋女畫家歐姬芙的足跡，孤獨的路途中除了歐姬芙外，就是拍照或將自己丟進攝影藝廊中，沈浸在攝影大師親自簽名的作品前，細細品味....原以為這麼長的一段旅程與際遇，應該好長一段時間不會再做那個攝影的夢.....

　　沒想到，流浪回來兩三週後，我又開始做著那個攝影的夢!

　　我的攝影是從【人】開始的...1999年的921地震，將我的經濟歸零，在這之前，我設計與銷售音響器材並且非常喜歡聆聽音樂，這是我的工作與收入，大地震將一切改變，因為經濟上完全崩潰而心情鬱鬱不樂，沒想到卻意外與知名美術評論家謝里法老師合作，拍攝三十位中部資深藝術家，後來又回故鄉嘉義，拍攝二十位美術家，至此，我對攝影的熱情一發不可收拾，又向銀行貸款了一筆龐大的金額，幾乎完全將工作拋至腦後而全心全意的進行一項艱鉅而沒有把握的拍攝計畫，這個計畫在拍攝的第一年就幾乎拍不下去，已經有放棄的心裡打算...

　　2004年元月，應國立歷史博物館館長黃光男先生的邀請，舉辦【台灣頭】人物攝影展，展出一百位台灣各個領域內最頂尖的人物，雖有掛萬漏一，但已經是那時的我盡全力所能完成的極限。

　　2005年，受邀與國立台灣交響樂團合作，配合該團六十週年慶，拍攝六十位知名的音樂家，並且集結成一本攝影集【六十，六十】。

　　由於【台灣頭】這個系列仍在拍攝中，連自己都覺得我的攝影好像脫離不了知名的人物肖像。

　　實際的情況當然不是如此，否則也不會作著那個攝影的夢。

　　從1996年底開始拍照自娛以來，我時常對著周遭的尋常人物，以及能觸動我內心的景物按下快門，1997年初在阿里山所拍攝的作品【冬之旅】是我【風景 安靜】系列作品的開始，而隔年所拍攝的【原住民兄弟】則是【平凡人物】系列的切入點，兩個系列都是當初在矇矇懂懂的情況下持續拍攝的作品，沒什麼計畫，也沒想到要拍出什麼成績。

　　2005年，我應台東縣政府文化局邀請，除參加他們舉辦的端午節詩人吟詩活動外，也受邀到蘭嶼去，由於從來不曾離家那麼久，內心非常寂寞，某個陰暗的黃昏，竟然在蘭嶼拍到【寂靜之聲】這張作品，那時的心情就如那個畫面，自此，才驚覺攝影於我是訴說著什麼樣的語言。

　　對蘭嶼的喜愛，引爆我體內所流著另一半屬於海島人的血液，在暌違二十五年之後，2005年八月，我重新踏上母親生長的島嶼-澎湖，距離母親過世已有17年之久...

　　2007年五月，受到女畫家歐姬芙的召喚，我來到美國新墨西哥州流浪，除了追逐女畫家走過的足跡外，也在尋找個人內心的寧靜。

　　【風景 安靜】是這11年來，日常生活以及孤獨旅程當中所拍下的作品。

　　一連串的攝影過程，受到多位人士的幫忙，當時音響的老師陳弘典；引領我進入人物攝影的謝里法老師；給我相當多鼓勵的雕刻大師朱銘；邀請我在國家藝廊舉辦展覽的前國立歷史博物館館長黃光男；邀請我拍攝音樂家群像的前國立台灣交響樂團蘇忠團長；攝影大師柯錫杰邀我帶他到嘉義去采風，讓我見識到攝影家的態度；給我信心的則是另一位攝影前輩張照堂，沒有他大力協助挑選作品以及為作品排序，攝影集與展覽不可能這麼順利，書名【風景 安靜】就是取自他作序的標題；對於這些與一些未提到的人物，我有說不盡的感謝之意。

　　我仍會做著那個攝影的夢，雖然內人說太神經質了，但是，唯有持續做著這樣的夢，我知道，對於攝影的澎湃熱情我仍然持續著。

I've still had a dream about taking photos

In 2005, I wrote a short article, "*I had a dream about taking photos*", as the preface for my photography book *Sixty years, sixty people*. It's about a dream I would usually had when I could not take photos for a while. In the dream, I was taking pictures. And then, a fantastic sight or a dramatic light came in, but I couldn't either find the right lens or load the film on time......then, I woke up in a shock!

In may 2007, I wandered in New Mexico, tried to follow the footsteps of painter Georgia O'Keeffee. On the solitary journey, I did nothing but taking photos and throwing myself into the photo galleries. I immersed myself into those originals with the masters' autograph, followed their shadows and lights... After the long journey and those unusual experience, once I thought the strange dream would disappear for a while...

However, it came back to me just two or three weeks after I came home.

I started to take photos on the subject of "people".

The "921 Earthquake" happened in 1999 took away all my properties. In the past, I enjoyed music very much. I lived by designing and selling stereo sets, until the earthquake changed my life. The sudden collapse of my originally stable income stroked me heavily. Unexpectedly I came across a chance to cooperate with the famous art critic, Mr. Shaih Li Fa, to photograph 30 famous artists located in the middle of Taiwan. After that I came back to my hometown Chayii to shoot other 20 artists. From then on, my enthusiasm for photography bloomed. I borrowed a huge sum of money from a bank. I threw my jobs away and dedicate myself to an uncertain project. At the first year, I could hardly keep going and wondered if I should give up.

In January 2004, the curator of National Museum of History, Mr. Huang Kuang-nan, invited me to hold an exhibition entitled *Face's Talk*. It was a series of portraits of one hundred of the preeminent people from different realms of Taiwan. I tried my best to complete the project, though I know there must be someone to be left out.

In 2005, I was invited by National Taiwan Symphony Orchestra to take photos of 60 famous musicians for celebration of their 60th anniversary. In the end, we put the portraits together and published a photography book, *Sixty years, sixty people*.

The portraits of *Face's Talk* still go on. It seems that my photos are always about celebrity.

But it's not true; otherwise the dream about taking photos should have disappeared.

When I started to take photos for fun in 1996, I always pointed my camera at people and objects of ordinary life. They touched me naturally, without specific purpose. "*Winterreise*" (1997) taken in Mt. Ali started the series of photos named *Voice inside*. In the next year, "*The Aboriginal Brothers*" was the starting point of the series named Nobody. Both of the series had no particular plans or expectation. I just followed my intuition.

In 2005, Cultural Affair Bureau of Taitung County Government invited me to a citing poetry event as part of Dragon Boat Festival. Afterwards I was invited to Orchid Island with them. I had never been away from home for such long time and I felt very lonely. One day in the dark evening, I took a photo later named "*The Sound of Silence*". The picture completely expressed how I felt at that time. Since that, I was surprised to find what and how much photography tries to talk to me.

The love to Orchid Island strongly stirs the blood inside me partially belonged to ocean and islands. In August 2005, after 25 years since my last visit, I came back to Penghu, the island of my mother's hometown. It has been 17 years since my mother passed away.

Evoked by the spirit of Georgia O'Keeffee, I went to New Mexico to follow her footsteps and to look for peace of my inner self.

Voice inside was a series of photos taken on my ordinary life and solitary journey in the past 11 years.

There has been great help to me from the seniors and precursors on my road of photography. I would like to pay my deepest gratitude to my teacher Mr. Chen Hung-Tien; Mr. Shaih Li Fa, who opened to me the door of portrait photography; the master of carving, Mr. Ju Ming, who encouraged me a lot; the pre-curator of National Museum of History, Mr. Huang Kuang-Nan, who invited me for an exhibition in the national gallery; and the master of photography, the pre-chief of National Taiwan Symphony Orchestra, Mr. Su Chung, who invited me to take portraits of musicians. Mr. Co Si-Chi, who took me on a field trip and showed me the way of a professional photographer. Another precursor of photographer I am really grateful is Mr. Chang Chao-Tang, who has had strong confidence on me and helped me with selecting works and putting them in good order. Without his great support, there wouldn't be such success with my photo books and exhibition. In fact, the title of my book *Voice inside* comes from the same title of preface written by him. To all these people I have and haven't mentioned, I have the greatest respect.

I still have the dream about taking photos. Although my wife regards it as a bit neurotic, the dream clearly tells me that my passion for photography will keep on burning.

Tseng Miin-Shyong

特別感謝攝影家　張照堂老師

目　錄

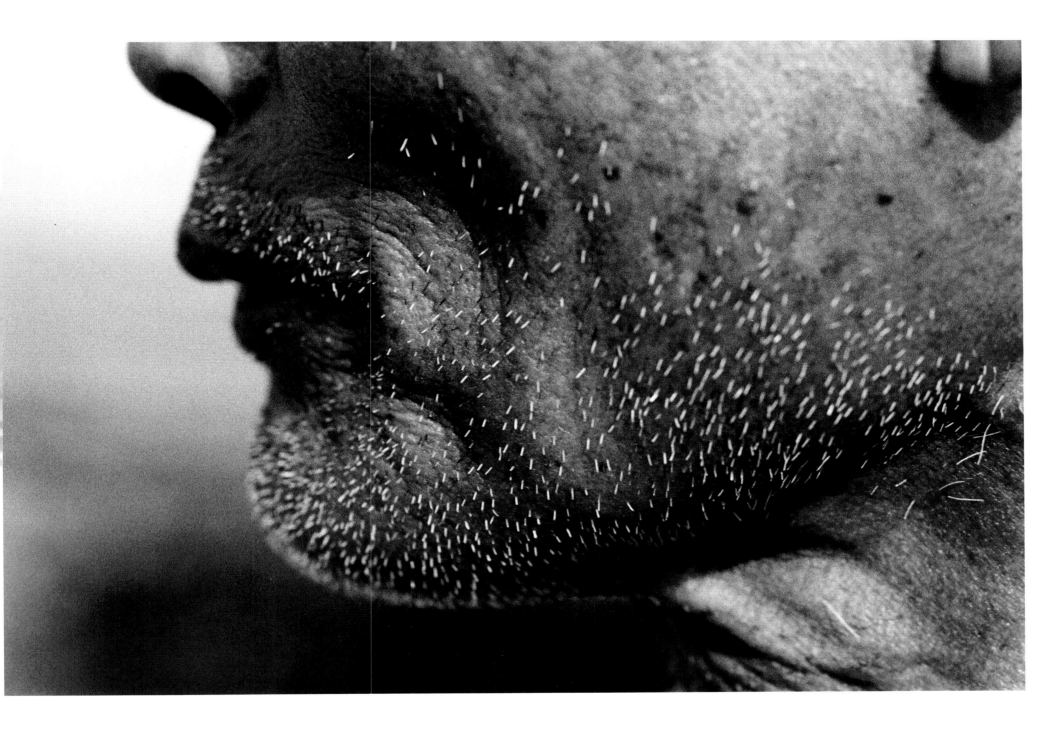

p79
美國 新墨西哥州 2007
Pueblo de Taos

p81
美國 新墨西哥州 2007
Ghost Ranch

p83
美國 新墨西哥州 2007
Ghost Ranch

p85
美國 新墨西哥州 2007
Ghost Ranch

p86
美國 新墨西哥州 2007
Unknow

p87
美國 新墨西哥州 2007
Ghost Ranch

p89
美國 新墨西哥州 2007
Unknow

p91
美國 新墨西哥州 2007
Abiquiu

p92
美國 新墨西哥州 2007
Ghost Ranch

p95
美國 新墨西哥州 2007
Santa Fe

p97
美國 新墨西哥州 2007
Santa Fe

p98
美國 新墨西哥州 2007
White Sand Dune

p99
美國 新墨西哥州 2007
White Sand Dune

p101
美國 新墨西哥州 2007
White Sand Dune

p103
美國 新墨西哥州 2007
White Sand Dune

p105
美國 新墨西哥州 2007
White Sand Dune

p107
美國 新墨西哥州 2007
White Sand Dune

p109
美國 新墨西哥州 2007
White Sand Dune

p110
美國 新墨西哥州 2007
Abiquiu

Self Portrait   2004   曾敏雄

曾敏雄

嘉義縣人，學生時代即著迷於音響設計。

當兵時，第一次寫相聲，獲得國軍文藝金獅獎佳作。

創辦 [鋅匠] 音響技術雜誌

成立 [音響種子] 工作室；並以 [Formosa] 為品牌，設計音響器材銷售

電台古典音樂主持人

1996年歲末開始拍照自娛

1999年921地震將其經濟歸零，卻意外接觸美術評論家謝里法，開啟拍攝人像生涯。

1999年歲末開始拍攝中部藝術家群像

2000年歲末拍攝嘉義地區美術家群像

2001年，受雕刻大師朱銘邀請，為其拍攝太極拱門作品。

2001年有了拍攝 [百年英雄] 計畫，由於很難取得被拍者資料，一度想放棄

2002年元月，[百年英雄] 通過國家文化藝術基金會的審核，獲得獎項補助，再度積極拍攝。

2003年完成 [百年英雄] 人物拍攝計畫；獲國立歷史博物館黃光男館長邀請，以 [台灣頭] 為名，舉辦展覽，並獲 EPSON 公司贊助影像輸出。

2003年開始拍攝台東地區藝術家群像。

2004年獲邀擔任台灣愛普生（EPSON）公司百萬攝影比賽系列講座講師。

2004年台東縣美展攝影評審；新社鄉攝影比賽評審。

2004年與國立台灣交響樂團合作，拍攝音樂界人物群像。同年第一次踏上蘭嶼，受到該島嶼特殊的人文風景之吸引，隔年再度前往。

2005年，因為蘭嶼的影響，引爆體內另一半屬於海島人的血液，在睽違25年之後，再度踏上母親生長的島嶼-澎湖。並開始以【母親的島嶼】為名，拍攝這塊島嶼上的人物與攝影家內心對母親的懷念。

2005年，再度受雕刻大師朱銘邀請，為其拍攝人間系列等五本作品集。

2006年受相機代理商邀請，為Nikon School 特約講師。

2006年二月，受攝影大師柯錫杰邀請，開車帶老攝影家至嘉義山區及海邊小村落采風；同月底至越南及柬普寨旅行，坐長途巴士越過邊境時，為柬普寨當地人民困苦的生活感到不捨；並獨自前往吳哥窟體驗千年古蹟的歷史。

2007年四月，受張照堂老師與劉振祥先生的邀請，與他們共同展出黃海岱老先生的人物肖像作品。

2007年五月，尋著美國女畫家歐姬芙的腳步，來到新墨西哥州流浪，深受墨西哥古老文化及當地沙漠地質的吸引，除了拍攝出極具個人代表性的作品外，並親身體驗世界一流攝影家的真跡作品。

2007年七月，有感於流浪時受到美國藝文界友人的肯定，決定整理出十一年來所拍攝的兩個系列作品【平凡人物】與【風景 安靜】，十月初，這兩個系列的作品受到前輩攝影家張照堂老師的肯定， 為其挑選作品舉辦展覽及出版個人攝影集。

出版

2004年元月　【 台灣頭-曾敏雄人物攝影集 】 - 國立歷史博物館發行
2004年元月　【 冬之旅 】 攝影集 - 曾敏雄工作室發行
2005年八月　【 中部百年美術史 】 - 台中市政府
2005年十月　【 六十‧六十 - 曾敏雄人像攝影集 】 國立台灣交響樂團發行。
2005年十二月 【 台灣頭 - 文化筆記書 】 東森媒體集團
2007年十一月 【 風景 安靜 】 曾敏雄工作室

展覽 (個展)

2000年九月　【 12分39秒 - 中部藝術家人像攝影展 】 --- 景薰樓藝廊
2002年五月　【 陳庭詩遺作暨人像紀念展 】 --- 國立台灣美術館
2004年元月　【 台灣頭 - 曾敏雄人物攝影展 】 --- 國立歷史博物館
2006年元月　【 台灣頭 】台灣愛普生公司藝廊。
2007年十二月 【 散步‧安靜 】 台北爵士攝影藝廊
2008年三月　【 風景 安靜 】 台灣國際視覺藝術中心

展覽 (聯展)

2004年　　【 台東縣美展 】
2007年　　【 黃海岱百年榮耀攝影展 】

典藏

【 陳庭詩 】 人像作品一幅 --- 國立台灣美術館。
【 台灣頭 】 人像作品十幅 --- 國立歷史博物館

Tseng Miin-Shyong

Born in Chayi County.

Have became addicted to audio design since high school.

During military service, earned a Golden Lion Award of Chinese Army Forces by writing the first lyric of Chinese talk show.

Published the Audio technology magazine named "Welder"

Established "Audio Seed" studio, and created the brand" Formosa" to design and sell audio equipments.

DJ , for classical music programs in a radio station.

1996 Started taking photographs for own pleasure.

1999 The "921 Earthquake" took away all his properties, but open a chance to cooperate with the art critic, Mr. Shaih Li Fa.Thus started Tseng's career of portrait photography.

1999 To photograph 30 artists located in the middle of Taiwan

2000 To photograph 20 artists located in Chayii.

2001 Invited by the master of carving, Mr. Ju Ming, to photograph his work ," Tai Chi"

2001 Began the project named "Centennial Heroes" . Considering the difficulties in data collection, once abandoned the project.

2002 "Centennial Heroes" was approved by the National Cultural ans Art Foundation for awarded and funded in January.

2003 "Centennial Heroes" was complete. Invited by Mr. Huang Kuang-Nan, the National Museum of History, to hold an exhibition entitled *Face's Talk* and EPSON company supported the image print output.

2003 To photograph artists located in Taitung.

2004 Lecturer in the series of Taiwan Epson Million Dollar Photography Competition

2004 Judge, Photography Section of the Taitung County Art Exhibition.

Judge, Hsin-Sher Township Photo Competition.

Invited by National Symphony Orchestra to take photos of famous musicians.

Visited Orchid Island for the first time. Attracted by the special customs and senery, he went here again next year.

2005 The love to Orchid Island stirred the blood inside Tseng partially belonged to ocean and islands.

After 25 years since his last visit, Tseng came back to Penghu and began to take photos named "Mother's Island" in memory of his mother.

Invited by the master of carving, Mr. Ju Ming again, to photograph his five books about "Living World Series"

2006 Lecturer in Nikon School.

2006 In February, drove with the master of photography, Mr. Co Si-Chi on a field trip around the villages of mountains and the seashore in Chayji.

In the same month, traveled to Viet Nam and Cambodia. Took a bus across the border. Felt sorrow about people's hardships there. Traveled along to experienced history of the great antiquities in Siemreap.

2007 In April, hold an exhibition with the predecessors Mr. Chang Chao-Tang and Mr. Liu Chen-Hsiang in memory of Mr. Hung Hai-Tai.

In May, followed the footsteps of American painter, Georgia O'Keeffee, Tseng wandered in New Mexico. He was attracted deeply by ancient Mexican culture and the landscapes of dessert; he not only took some represetive photos but also experenced the world of the photography masters' originals.

In July, for the affirmation of American friends in the arts, made up his mind to arrange the works took in the past 11 years into two series. Named them as *Nobody* and *Voice Inside*.

In October, the predecessor of photographer, Mr. Chang Chao-Tang, put confidence in the two series, and helped to select works for exhibitions and photo books.

Pubications

2004 *Face's Talk* - Portrait photography by Tseng Miin-Shyong , by National Museum of History.

2004 *Winterreise*, by Tseng Miin-Shyong Studio.

2005 *History of Arts in the middle of Taiwan*, by Taichung City Government.

2005 *Sixty, sixty - Portrait photography by Tseng Miin-Shyong* , by the National Taiwan Symphony Orchestra.

2005 *Face's Talk* - a culture notenook, by Eastern Group.

2007 *Voice inside*, by Tseng Miin-Shyong Studio.

Solo Exhibitions

2000 *12 Minutes 39 Seconds - Portrait Photograph of Artists in the middle of Taiwan*, in Ching-Shiun Lo Gallery.

2002 *Ting Shih Chen Memorial Exhibition*, in National Taiwan Museum of Art.

2004 *Face's Talk* - Portrait photography by Tseng Miin-Shyong, in National Museum of History.

2006 *Face's Talk*, in the Gallery of Taiwan Epson.

2007 *Walking Inside*, in Jazz Imagine Gallery.

2008 *Voice Inside*, in TIVAC.

Group exhibitions

2004 The Arts in Taitung County.

2007 *One hundred years of Honor, Hung Hai-Tai Memorial Exhibition.*

Collections

"*Ting Shih Chen*" Portrait, by National Taiwan Museum of Art.

Face's Talk - 10 pieces of Portraits , by National Museum of History.

國家圖書館出版品預行編目資料

風景 安靜：曾敏雄攝影集 / 張照堂編輯. --
　　初版. -- 臺中市：曾敏雄工作室，　民96.12
　　　面；　公分

　　ISBN 978-986-83915-0-5（平裝）

　　1. 攝影集　2. 風景攝影

957.2　　　　　　　　　　　　　　96022011

風景 安靜　曾敏雄攝影集

作　　　者	曾敏雄
編 輯 指 導	張照堂
執 行 編 輯	辜素琴
美 術 編 輯	蘭耀先　陳彥錞
翻　　　譯	陳韻如

發 行 人	曾敏雄
出 版 單 位	曾敏雄工作室
	地址：台中市西區五權西三街89號2樓
	電話：04-23785535
	E-mail：audioseed@yahoo.com.tw

印　　　刷	飛燕印刷有限公司
	台北縣中和市橋安街17號6樓
	電話：02-22476705

出 版 日 期	中華民國96年12月
版　　　次	初版
定　　　價	新臺幣900元整

I S B N	978-986-83915-0-5（平裝）